Brent Jarvis

General Keplerian Dynamics (GKD). 2nd Edition

A testable unified model of the Universe

GRIN Verlag

Bibliografische Information der Deutschen Nationalbibliothek:

Die Deutsche Bibliothek verzeichnet diese Publikation in der Deutschen National-bibliografie; detaillierte bibliografische Daten sind im Internet über http://dnb.d-nb.de/ abrufbar.

Imprint:

Copyright © 2014 GRIN Verlag GmbH
Druck und Bindung: Books on Demand GmbH, Norderstedt Germany
ISBN: 978-3-656-64292-3

This book at GRIN:

http://www.grin.com/en/e-book/271676/general-keplerian-dynamics-gkd-2nd-edition

GRIN - Your knowledge has value

Der GRIN Verlag publiziert seit 1998 wissenschaftliche Arbeiten von Studenten, Hochschullehrern und anderen Akademikern als eBook und gedrucktes Buch. Die Verlagswebsite www.grin.com ist die ideale Plattform zur Veröffentlichung von Hausarbeiten, Abschlussarbeiten, wissenschaftlichen Aufsätzen, Dissertationen und Fachbüchern.

Visit us on the internet:

http://www.grin.com/

http://www.facebook.com/grincom

http://www.twitter.com/grin_com

General Keplerian Dynamics (GKD) 2nd Edition
A testable unified model of the Universe

Brent Jarvis
Kennesaw State University

Abstract

Newton generalized Kepler's laws of planetary motion when he developed his laws of universal gravitation. Modifications to Newton's generalizations are submitted which offer a novel solution for the galaxy rotation problem and a unified model of the Universe. Observable evidence and experimental predictions are also submitted which can prove the unified model.

Introduction

This paper will outline the basics of General Keplerian Dynamics (GKD) and discloses predictions which can prove the Lorentz-Mandelbrot Fractal Electrodynamic Astronomical Model (FLAME) illustrated below. To the author's knowlege, FLAME is the **only** model that can either be proven or disproven experimentally (unlike string theory, M-theory, etcetera). I leave it up to the experimental physicists to test the novel model.

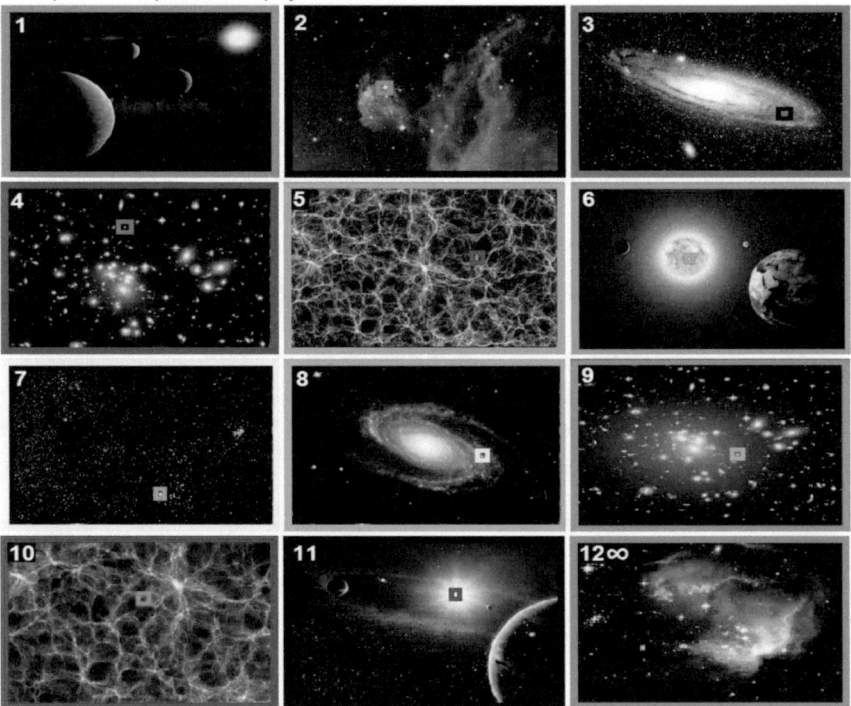

FIG. 1: The Lorentz-Mandelbrot Fractal Electrodynamic Astronomical Model (FLAME). The grey frame 1 zooms out to the black frame 2, which zooms out to the magenta frame 3, etcetera.

In honor of Einstein, let's assume there is an inertial reference frame **RF(A)** established for a train at rest and the position of a point **P** on the train is located at the origin of **RF(A)**. While the position of **P** is static at the origin of **RF(A)**, the velocity of **P** relative to a reference frame established for the Milky Way **RF(B)** is ≈ **792,000 kph**. The velocity of **P** is **0** when its position is certain in **RF(A)**, but in order to coordinate its exact position after an arbitrary amount of time in **RF(B)** it would have to be determined with a wave function (analogous to Heisenberg's uncertainty principle). The angular momenta of celestial systems are therefore dependent upon the volume of the inertial coordinate system. An excellent resource for scale dependent celestial mechanics is Laurent Nottale's theory of Scale Relativity[1].

A temporal phenomenon also emerges from the scale dependency of angular momentum since, according to the Lorentz transformation equation:

$$[1]\ t' = t - (vx\,/\,c^2)\,/\,\sqrt{1 - (v^2/\,c^2)},$$

time is also dependent upon the volume of the inertial coordinate system. This would explain the incompatibility between general relativity and quanutm mechanics since angular momentum, time, and therefore gravity, are all scale dependent. This can be tested experimentally via the FLAME model in **FIG. 1**. According to special relativity, the velocity of light in a vaccum **c** is constant independent upon the scale of an inertial frame of reference. If time is scale dependent, and **c** is not scale dependent, then the effects of quantum entaglement will be be limited to **c²** relative to our frame of reference (the fractal radiation of **frame 1** in **FIG. 1** would propagate at **c²** relative to **frame 6** and the radiation of **frame 11** would be observed as cosmic microwave background radiation). Recursive fractal radiation can also explain zero point energy / dark energy and fractal matter can explain dark matter. Of the four experimental predictions that will be disclosed in this paper, this is the most important, and fortunately we are on the brink of testing quantum entanglement effects at scales comperable to prove or disprove FLAME[2].

A generalized 1st law of planetary motion (GKD1) will also be submitted which synthesizes the conic path of a secondary with its orbital inclination via a non-inertial toroidal reference frame embedded within an inertial Cartesian coordinate system.

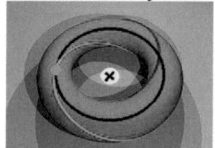

FIG. 2: Image is exaggerated. The center of mass (COM) is not positioned at one of two possible foci, but at the origin of the toroidal cavity (represented by the "x"). The torus can be constructed from an auxiliary circle with a radius equivalent to the secondary's apoapsis distance and a minor auxiliary circle with a radius equivalent to the distance of its periapsis (relative to the COM). The radius of the coplanar orbital axis (black) is equivalent to the secondary's average true anomaly distance (equivalent to the semi-minor axis of a Keplerian ellipse). The yellow and brown dots represent a primary and secondary respectively. The blue and green paths are Villarceau circles (example orbital paths of a secondary).

GKD1 can be stated as: A secondary's geodesic is embedded within a degenerative fractal horn torus with the center of mass at the origin of the toroidal cavity.

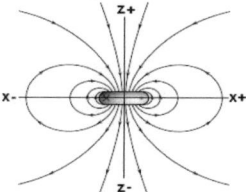

FIG. 3: An example half section of a degenerative fractal horn torus. Circular, elliptical, parabolic, and hyperbolic trajectories are embedded within each quadrant of the torus along the z-axis when the coplanar orbital axis is oriented along the x and y plane.

It will be shown within the generalized 2nd law of planetary motion (GKD2) that the orbit of a secondary has three primary axes of rotation. I refer to the three axes as the major, minor, and torque axes.

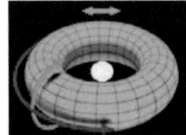

FIG. 4: Image is exaggerated. The major axis is represented by the blue vector, the minor axis is represented by the red vector, and the green pseudovector represents the torque axis.

Even though an ellipse and a Villarceau circle are geometrically equivalent relative to a two dimensional reference frame there is a subtle difference between the two in three dimensions. While an ellipse and a Villarceau circle both trace the oscillatory cycle of a secondary's apoapsis and periapsis relative to the COM, a Villarceau circle also traces an additional cycle which is perpendicular to the coplanar orbital axis. The additional cycle contains two reference points that I refer to as the crest and trough, in which the crest is perpendicular to the coplanar axis in the northernmost polar direction. The distance between the crest or trough from the coplanar axis will be referred to as the orbital amplitude.

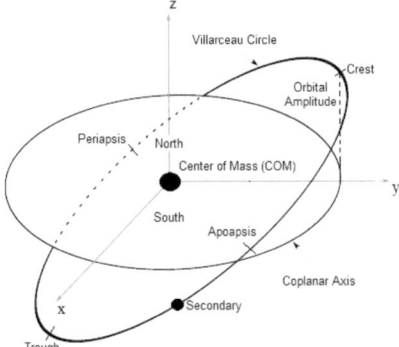

FIG 5: Image is exaggerated. An illustration of the orbital amplitude and the poloidal points of reference.

GKD3 will then show there are two periods that must be considered in Newton's version of Kepler's 3rd law instead of only one period. Villarceau trajectories result from a **1:1** ratio between a secondary's major and minor periods (which I refer to as its polar frequency). The introduction of the polar frequency term procures orbital paths that are not limited to conic sections. Geological evidence[3] indicates a **4:1** polar frequency for our solar system's orbit.

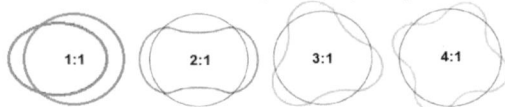

FIG. 6: Stellar orbital paths relative to galactic nuclei are not limited to conic sections according to the formalism of GKD3.

A Generalized 2nd Law of Planetary Motion (GKD2)

Kepler's second law of planetary motion can be stated as: A line joining a planet and the Sun sweeps out equal areas during equal intervals of time.

GKD2(a) can be stated as: The radii of a secondary's major (r_1), minor (r_2), and torque (r_3) axes, joining the major, minor, and torque points respectively, individually sweep out equal sectors during equal intervals of time (GKD2(a) will be expanded upon).

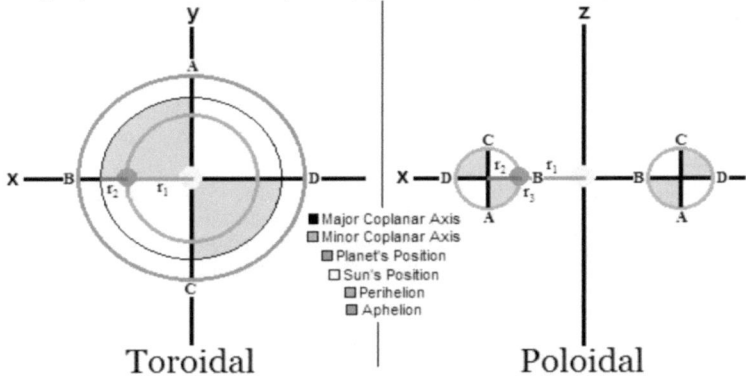

Major Coplanar Axis
Minor Coplanar Axis
Planet's Position
Sun's Position
Perihelion
Aphelion

Toroidal   Poloidal

FIG. 7: Image is exaggerated. The radii r_1 and r_2 are overlapping since the planet is at perihelion. The radius of the torque axis (r_3) is miniscule. The planet orbits a perpendicular fractal torus along the minor coplanar axis which is observed as precession (represented by the torque axis pseudovector in FIG. 4).

GKD2(b) can be stated as: The radius of a secondary's major axis (r_1) is equivalent to the semi-minor axis of its Keplerian ellipse when unperturbed by outside forces, or

$$[2] \ r_1 = b = a\sqrt{1-e^2},$$

where **b** is the semi-minor axis, **a** is the semi-major axis, and **e** is the eccentricity of the ellipse. The radius (r_1) is also equivalent to the geometric mean of the secondary's apoapsis and periapsis distances relative to the major point (barycenter), or

$$[3] \ r_1 = \sqrt{r_{min} r_{max}}.$$

GKD2(c) can be stated as: The angle of the secondary's toroidal major axis, relative to any other plane of reference, can be determined from the angle of r_1 relative to the plane of reference when the secondary is at crest or trough (when r_2 is perpendicular to r_1 and the hypotenuse is joined by the major and torque points).

GKD2(d) can be stated as: The relationship between the major axis revolution period, the minor axis revolution period, and the polar frequency of a secondary's orbit is:

$$J = P_M / P_m$$

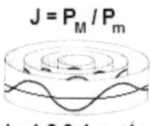

$$J = 1, 2, 3, 4, n, \text{etc.}$$

where P_M is the major period, P_m is the minor period, and J is the polar frequency. Additional standing wave formulas are:

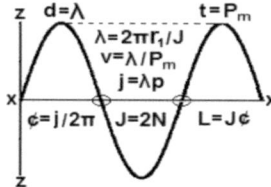

where λ is wavelength, v is velocity, j is an astronomical analog of Planck's proportionality constant, ϕ is the reduced constant, p is momentum (where mass is measured in Solar Mass units), L is angular momentum, N is the node quantity, d is distance in light units, and t is time in terrestrial minor periods (P_{Em}). Considering only the Sun-Earth system, the radius of the Earth's major axis (r_{E1}) relative to the Sun's COM is

[4] $r_{E1} = \sqrt{r_{min} r_{max}} \approx \sqrt{(0.9832898912 \text{ AU}) (1.0167103335 \text{ AU})} \approx 0.999860486872609 \text{ AU}$,

where **AU** is an astronomical unit. The barycenter distance can be determined by:

[5] $r_{E1} = a / (1 + m_1 / m_2)$,

where a is the distance derived from equation [4] and m_1 and m_2 are each of their masses. Compensating for the barycenter we get $r_{E1} \approx 0.99986015 \text{ AU} \approx 149{,}576{,}950{,}315$ meters, which can be converted into light minutes by:

[6] $r_{E1} / 1 \text{ lm} \approx 8.315583339808087 \text{ lm}$.

Since the node quantity is **2** for the Earth's orbit ($J = 1$), the Earth's wavelength λ_E is

[7] $\lambda_E = 2\pi r_{E1} / 2(0.5) \approx 52.24835106131297 \text{ lm}$.

The Earth's standing wave velocity (not its velocity relative to the Sun's position) can be determined from:

[8] $v = \lambda_E / P_m \approx 52.24835106131297 \text{ lm} / P_{Em}$,

where P_{Em} is the Earth's minor period. The velocity can then be used to deduce the astronomical analog of Planck's proportionality constant, which is approximately (on a stellar scale not a galactic scale):

$$[9] \, j = \lambda p \approx 9.4124100318113 \cdot 10^{-3} \, \text{ls}^2 \cdot M \odot / P_{Em}.$$

The reduced constant is simply $j \, / \, 2\pi = ¢$. If the predictions that will be submitted confirm GKD, the angular momenta of stellar systems are quantized to:

$$[10] \, L = J¢.$$

It is important to note that the minor period is measured by the amount of time that elapses between two passages of a crest, or two passages of a trough. Since the toroidal major axis radius is less than the Keplerian semi-major axis the standing wave velocity of a secondary is less than the mean velocity calculated with Kepler's version, and the waves are not precisely sinusoidal for reasons that will become apparent in a moment.

GKD2(e) can be stated as: The minor axis is perpendicular to the major axis, and the torque axis is perpendicular to the minor axis, when unperturbed by outside forces. Since this is analogous to eddy currents, I refer to GKD2(e) as the eddy effect. Fractal iteration of the eddy effect is limited by the secondary's intrinsic center of mass point.

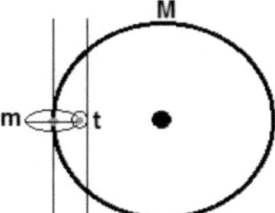

FIG. 8: Image is exaggerated. The torque axis (t) would be miniscule. The minor axis (m) would appear as a line in 2-D. The "M" represents the major axis. The black, green, and red dots represent the major, minor, and torque points respectively.

GKD2(f) can be stated as: The curl of the eddy effect obeys the right-hand rule for vector cross products relative to the secondary's "mass current" along the major axis. The rotation vector of the minor axis has reflection symmetry along the z-axis of an inertial Cartesian coordinate system when the major axis is oriented on the x and y plane. The pseudovector of the torque axis is mathematically equivalent to a 3-D bivector.

GKD2(g) can be stated as: In accordance with Newton's 3rd law of motion, GKD2(a-f) are also true for the primary body's orbit. The major point is equivalent to the barycenter.

I have consolidated GKD2(a-g) into what I refer to as gyrographs, which simplify calculations pertaining to the position and/or momentum of a secondary at any given time in its orbit. I refer to the diagrams as gyrographs since all of the terms that pertain to the angular momentum of a secondary are kept constant except for the relative gyration of each axis of rotation within an inertial Cartesian coordinate system (see **FIG. 9** on the next page).

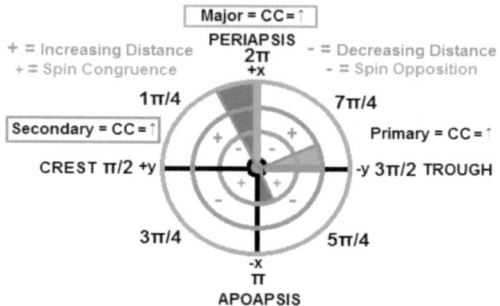

J=1; m = constant; v = constant; $r_{(1,2,3)}$ = constant; sin θ = constant.

FIG. 9: Image is exaggerated. The axes of rotation are superimposed on the same plane but the gyrograph can be disassembled to determine the 3-D position of the secondary by linking the radii back together in their proper orientation (+z is coming out of the page). The radius of the major axis (blue) always defines the "time" and all three radii "tick" counter-clockwise (CC=↑) at the same rate since J=1. The radius r_2 is 90° out of phase with r_1 and r_3 is 90° out of phase with r_2, 180° out of phase with r_1. The intrinsic spin of the primary and secondary are both CC=↑. The rotation vector of the minor axis (red) flips signs at periapsis and apoapsis (the "+" and "-" represent the increasing (→) and decreasing (←) distance of the secondary's position relative to the primary's position respectively). The pseudovector of the torque axis (green) flips signs at crest and trough (the "+" and "-" in this case represent rotation congruence (↑) and opposition (↓) relative to the secondary's intrinsic spin respectively).

Throughout a full revolution of a gyrograph we can see that the gyration of the torque axis is equivalent to the total intrinsic spin angular momentum of an electron (which can be treated as a point particle):

$$[11] \; S = ℏ \sqrt{½(½+1)},$$

where **S** is the spin angular momentum and **ℏ** is the reduced Planck constant:

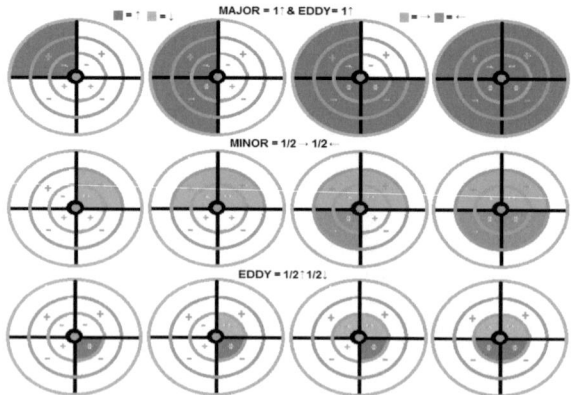

FIG. 10: A full revolution of the individual axes of rotation in a gyrograph. By precessing the gyrograph by ½↓1½↑=$2πr_3$ per wavelength the secondary's 3-D position can be determined at any time.

8

The gyration of the torque axis is ½↓ and 1½↑ per wavelength. The ½↓ should not be interpreted as an alteration in the secondary's intrinsic spin orientation, but as the frame-drag (torque) per wavelength due to the eddy effect. With this interpretation, the perturbations of a secondary's orbit predominately originate from the parity inversion of the torque axis every ½ revolution of the minor axis (represented by the torque axis pseudovector). There are also two types of torque that can be deduced from a gyrograph; a "boost torque" (magenta) and a "tranquil torque" (orange).

GKD2(h) can be stated as: In addition to Newton's generalization of Kepler's 2nd law, which is the conservation of angular momentum, torque is also conserved (experiments regarding torque conservation will be discussed next).

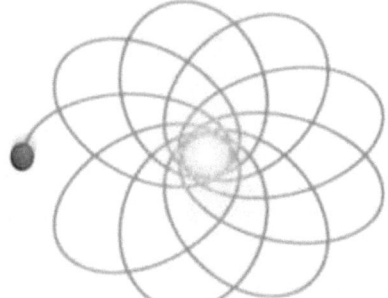

FIG. 11: Image is exaggerated. If L = mvr sin θ = constant then precession should not be observed. Assuming torque conservation (ΔL / Δt) amends the law of conservation of angular momentum.

Since time is intrinsically related to the Earth's rotation there is an obvious correlation between GKD and Einstein's theory of general relativity (GR). The primary difference between the two is the precession rate of an ↑ primary ↑ secondary system (such as the Sun-Earth system) relative to an ↑ primary ↓ secondary system (such as the Sun-Venus system). To the author's knowledge, there has yet to be an experiment that has tested the aphelion precession rate of Venus relative to its conventional perihelion precession. GR predicts a perihelion precession of ≈ **8.6** arc seconds per century, but the observed precession is ≈ **2.04** arc seconds per century. According to GKD, the "reverse" precession of Venus is ≈ **6.12** arc seconds per century (if the tranquil torque is **2.04** then the boost torque is **6.12**).

Since there are two known spin orientations for an electron (↑ or ↓), a method to test GKD on an atomic scale may be to repeat the Hafele-Keating experiment[4] with two atomic clocks onboard the same aircraft (one clock could have an ↑ proton, ↑ electron hyrogen-1 (protium) orientation, and the other an ↑ proton, ↓ electron protium orientation). Even though each clock will travel at the same relative velocity there may be a miniscule difference in their precession rates (assuming the spin orientations are kept constant and the aircraft maintains an opposite flight path (for optimal results) relative to the Earth's spin). Taking the effects of GR into account, the greater the velocity of the aircraft and the lower its altitude, the greater the assumed difference between the measured times on each atomic clock.

A Generalized 3rd Law of Planetary Motion (GKD3)

According to Kepler's third law of planetary motion, the square of a secondary's orbital period is proportional to the cube of the semi-major axis of its orbit. Newton later modified this law to include the mass of each body:

$$[12] \ m_1 + m_2 = A^3 / P^2,$$

where m_1 and m_2 are each of their masses (in Solar Mass units), **A** represents the semi-major axis distance (in Astronomical Units), and **P** represents the period of the orbit (in Years). As was discussed previously, however, there are at least two periods which must be considered in Newton's version of Kepler's third law. Until the predictions that were discussed previously can be verified experimentally, it will be assumed equation [12] can be modified to:

$$[13] \ m_1 + m_2 = A^3 / JP_M^2,$$

When **J=1**, equation [13] is equivalent to Newton's generalization, but introducing the polar frequency term geometrically enables orbital paths that are not limited to conic sections. With this interpretation, the consistency of a star's rotation speed independent of its distance from the galactic nucleus would be expected, so long as the value of **J** increases as the star's distance increases (which can also be experimentally investigated). Current estimates for our solar system's major period range from **225-250 Myr**, and geological evidence indicates a minor period of ≈ **60 ± 2 Myr**, from which a polar frequency of **4** for our system may be deduced[3]. A rough estimate of the observable mass within our radius would therefore be:

$$[14] \ m_1 + m_2 \approx 1,708,860,759.5^3 / 4(240,000,000)^2 \approx 2.17 \times 10^{10} \ M_\odot,$$

which would eliminate the necessity of dark matter.

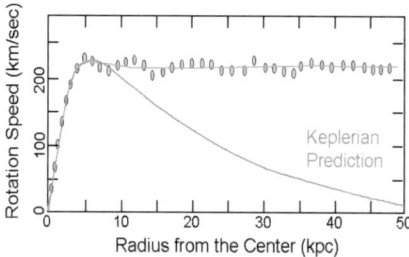

FIG. 12: Assuming J>1 as a star's distance increases eliminates the necessity of dark matter.

As discussed in the introduction, angular momentum is scale depenedent. While the formula **L=J¢** is efficient for planetary systems, the angular momenta of stellar systems relative to galactic nuclei are more efficiently determined by the stellar forumla **L=J¢_s**, where

$$[15] \ ¢_s = j_s / 2\pi = \lambda_s p_s / 2\pi,$$

where λ_s is the stellar wavelength and p_s is the stellar momentum.

10

Acknowledgements

This paper is dedicated to Cindy Lett. It would not have been possible without her.

References within paper:

[1] L. Nottale, "*Scale Relativity and Schrodinger's Equation*," Chaos, Solitons, and Fractals 9 (7): 1051-1061 (1998).

[2] X. Ma, T. Herbst, T. Scheidl, D. Wang, S. Kropatschek, W. Naylor, A. Mech, B. Wittmann, J. Kofler, E. Anisimova, V. Makarov, T. Jennewein, R. Ursin, and A. Zeilinger, "*Quantum teleportation using active feed-forward between two Canary Islands*," arXiv:1205.3909v1 [quant-ph] (2012).

[3] M. R. Rampino and R. B. Stothers, "*Terrestrial mass extinctions, cometary impacts, and the Sun's motion perpendicular to the galactic plane*," Nature 308, 709 – 712 (1984).

[4] J. C. Hafele and R. E. Keating, "*Around-the-World atomic clocks: Predicted Relativistic Time Gains.*" Science 177 (4044): 166-168 (1972).

Important additional references:

[5] H. Alfvén, "*On hierarchical cosmology*," Astrophysics and Space Science (ISSN 0004-640X), vol. 89, no. 2, January 1983, p. 313-324 (1983).

[6] A. L. Peratt, "*Introduction to Plasma Astrophysics and Cosmology*" Astrophysics and Space Science, v. 227, p. 3-1 (1995).

[7] H. S. Kragh, "*Cosmology and Controversy: The Historical Development of Two Theories of the Universe*," Princeton University Press, 488 pages, ISBN 0-691-00546-X (pp.482-483) (1996).

[8] O. Klein, "*Arguments concerning relativity and cosmology*," Science 171, 339 (1971).

[9] A. L. Peratt, "*Plasma Cosmology: Part I, Interpretations of a Visible Universe*," World & I, vol. 8, pp. 294-301, August (1989).

[10] A. L. Peratt, "*Plasma Cosmology:Part II, The Universe is a Sea of Electrically Charged Particles*," World & I, vol. 9, pp. 306-317, (1989).

[11] W. Thornhill, "Science's Looming 'Tipping Point'," posted on holoscience.com (2012).

[12] D. E. Scott, "*Magnetic Fields of Birkeland Currents*," posted on electric-cosmos.org.

[13] B. Mandelbrot, "*The fractal geometry of nature*," Macmillan, ISBN 978-0-7167-1186-5 (1983).